FROM **PEANUT**
TO **PEANUT BUTTER**

by Robin Nelson

Lerner Publications Company / Minneapolis

Lerner Publications Company
A division of Lerner Publishing Group
241 First Avenue North
Minneapolis, MN 55401 U.S.A.

Website address: www.lernerbooks.com

Library of Congress Cataloging-in-Publication Data

Nelson, Robin, 1971–
    From peanut to peanut butter / by Robin Nelson.
        p.   cm. — (Start to finish)
    Includes index.
    Summary: Briefly introduces the process by which peanut butter is made from peanuts.
    ISBN: 0–8225–0944–X (lib. bdg. : alk. paper)
    1. Peanut butter—Juvenile literature. 2. Peanuts—Processing—Juvenile literature. [1. Peanut butter. 2. Peanuts.] I. Title. II. Start to finish (Minneapolis, Minn.)
TP438.P4N45 2004
641.3'56596—dc21   2002153395

Manufactured in the United States of America
1 2 3 4 5 6 – DP – 09 08 07 06 05 04

Table of Contents

Peanut butter tastes good!

How is it made?

A farmer grows peanuts.

A farmer plants peanut seeds. The seeds grow into peanut plants. Little buds called **pegs** grow on the plants. The pegs push into the ground and grow under the dirt. The pegs become peanuts.

The sun dries the peanuts.

The farmer uses a digging machine to dig up the peanut plants. The peanuts are wet and dirty. A shaking machine shakes the dirt off the peanuts. The peanut plants are spread on the ground to dry in the sun for a few days.

The peanuts are picked up.

A farmer uses a machine called a **combine** to pick up the peanut plants. The combine puts the peanuts in a large box. Then it pulls the peanuts off of the plants.

The peanuts are sorted.

Trucks take the peanuts to a **shelling factory**. A shelling factory is a place where peanuts are sorted and cleaned. Workers sort the peanuts by size. Small peanuts are used to make peanut butter.

11

Machines remove the shells.

A shelling machine pushes the peanuts through small holes in a piece of metal. The shells are crushed. A fan blows the shells away.

13

Workers check the peanuts.

Workers check the peanuts. Peanuts that are not good to eat are removed. The good peanuts are poured into bags. The bags are sent to a factory to be made into peanut butter.

The peanuts are cooked.

The peanuts are put into big ovens. The ovens cook the peanuts. Cooking the peanuts makes them taste better.

17

Machines mix the peanut butter.

Machines crush the peanuts into very small pieces. Vegetable oil and salt are mixed in. Sometimes sugar or honey is added. Machines mix the peanut butter until it is smooth.

The peanut butter is put into jars.

A machine called a **filler** squirts the peanut butter into empty jars. Another machine puts lids on the jars. Trucks take the peanut butter to stores to be sold.

21

Peanut butter makes a great snack.

You can spread peanut butter on bread or eat it with celery. You can even eat it right out of the jar!

Glossary

combine (KAHM-byn): a machine that collects and cleans peanuts

filler (FIHL-ur): a machine that squirts peanut butter into jars

pegs (PEHGZ): the buds of a peanut plant

shelling factory (SHEHL-ing FAK-tur-ee): a place where peanuts are cleaned and sorted

Index